HOW TO THINK LIKE AN ENGINEER

The Lost Book of Creative Thinking for Youngsters

Dylan Austin Jr.

Contents

INTRODUCTION
Your Biggest Dream Awaits Fulfilment
What Engineers Do
How Engineers Solve Problems
Why Thinking Like an Engineer is Cool

CHAPTER 1:
The Engineer's Toolbox
Homework Time
Working Together to Create Something Amazing
Homework Time

CHAPTER 2:
The Engineering Design Recipes

CHAPTER 3:
Types of Engineers

CHAPTER 4:
Real-Life Engineering Challenges

CHAPTER 5:
Becoming an Engineer
What You Can Do Now
How to Stay Curious and Keep Learning
Cool Engineering Clubs and Competitions
Homework
Your Engineering Action Plan

CHAPTER 6:
Wrapping It Up: You're an Engineer
How to Keep Thinking Like an Engineer
Where to Go from Here
Homework: Your Engineering Reflection

Introduction

Your Biggest Dream Awaits Fulfilment

So, you're curious about engineering, huh? That's awesome, because let me tell you—a whole world of creativity, problem-solving, and cool inventions is waiting for you. Whether you're someone who loves taking things apart to see how they work, or you've got big ideas for making the world a better place, this book is your ticket to discovering just how incredible the life of an engineer can be.

But before we dive in, let's talk about what engineering actually is. You've probably heard the word before, but maybe you're not entirely sure what it means. Well, in the simplest terms, engineering is all about solving problems. In a more concise sense, Engineers are like detectives who use science, math, and a whole lot of creativity to figure out how to make things work better, faster, or smarter. They're the people behind the bridges we drive on, the gadgets we use every day, the clean energy we're starting to rely on, and even the space missions that take us to the stars.

Now, I know what you might be thinking: "Isn't engineering just for super smart people or math wizards?" Absolutely not! Here's the thing: if you're curious, if you like to ask questions, and if you're willing to try new things and sometimes fail along the way, you've already got what it takes to think like an engineer. You don't need to be a math genius or have a PhD to get started. All you need is a bit of imagination and a willingness to explore.

In this book, we're going to break down what it really means to be an engineer, and don't worry—we'll keep it fun and easy to understand. We'll explore different types of engineers, from those who design cool gadgets to those who build massive bridges or even work on projects that could help save the planet. We'll take you step by step through the engineering design process, where you'll learn how to turn your wildest ideas into real, working projects. And yes, you'll get your hands dirty with some awesome challenges that you can try at home, school, or wherever you like to tinker and create.

But more than just teaching you about engineering, this book is here to show you that thinking like an engineer is a superpower in itself. It's a way of looking at the world that helps you tackle problems head-on, dream up new ideas, and work with others to make those dreams a reality. Whether you want to invent the next big thing, help solve global issues like climate change, or just understand how the world around you works a little better, this book is your starting point.

So, buckle up and get ready for a journey filled with discovery, creativity, and plenty of "aha!" moments. By the time you're done reading, you'll not only understand what engineers do, but you'll also be ready to start thinking—and acting—like one. And who knows? The next great invention or world-changing idea might just come from you.

Are you ready to dive in? Let's get started!

WHAT ENGINEERS DO

Have you ever thought about what engineers actually do? It's way more than just tinkering with gadgets or wearing hard hats. Imagine this: everything around you, from the bed you sleep in to the tablet you play games on, has a little bit of engineering magic in it. Engineers are like the world's problem-solvers and inventors all rolled into one!

Let's start with your morning. When you brush your teeth, the toothbrush and the toothpaste were designed by engineers. The water you drink from the tap? Engineers worked hard to make sure it's clean and safe. Even your favorite cereal box might have been designed by an engineer to keep the cereal fresh and crunchy. Pretty cool, right?

Now, think about going outside to play. You see buildings, bridges, cars zooming by, and maybe even an airplane flying overhead. All of these things were made possible by engineers. They figure out how to make things strong enough to hold up in bad weather, fast enough to get you where you need to go, and safe enough so that you don't have to worry about a thing. Imagine if you were in charge of designing the tallest skyscraper in the world. How would you make sure it doesn't fall over? That's the kind of challenge engineers take on every day!

But engineers don't just stop at building things. They also work on making the world a better place. Let's say there's a problem with too much trash piling up. An engineer might come up with a new way to recycle or even design a machine that turns trash into something useful. Or maybe there's a new game you love playing. Behind that game are software engineers who wrote the code to make it fun and challenging.

Have you ever played with LEGOs or built a fort out of pillows? If so, you've already taken the first steps toward thinking like an engineer. It's all about using your imagination to create something new. And just like you, engineers love to build and create. They take their ideas and turn them into real-life things that help people every day.

Let's pretend you're an engineer for a minute. Imagine you've been asked to design a new playground for your school. What would you include? Maybe a super-fast slide, a giant jungle gym, or even a zip line! As an engineer, you'd draw out your ideas, think about how to make everything safe, and then work with a team to build it. The best part? When it's done, you'd get to see all your friends playing on something you helped create. How awesome is that?

So, what do engineers do? They imagine, create, and solve problems to make the world a better place. They turn dreams into reality, whether it's designing the next big toy, building a rocket to explore space, or finding new ways to protect our planet. And here's the coolest part: you can be an engineer too! All you need is a curious mind, a love for building and creating, and a passion for helping others. So keep dreaming, keep building, and who knows? Maybe one day, you'll be the one designing the next big thing that changes the world!

How Engineers Solve Problems

Let's say you're trying to build the perfect treehouse. You have the wood, nails, and a hammer, but there's one problem: the tree you chose isn't very sturdy. What do you do? Do you give up? No way! You start thinking like an engineer. Maybe you could add extra support beams or pick a different tree. Engineers never just quit when they hit a bump in the road—they get creative and find a way around it.

When engineers face a problem, they follow a special process called the Engineering Design Process. It's like having a step-by-step guide to solving any challenge. First, they Ask questions to understand the problem better. Then they Imagine different solutions, like brainstorming ideas with friends. Next, they Plan by choosing the best idea and figuring out how to make it work. After that, they Create a prototype, which is like a practice version of their idea. Finally, they Improve it by testing and making changes until it's just right.

Let's imagine you're trying to invent a new way to carry all your school books so you don't have to lug them around in a heavy backpack. First, you'd Ask yourself: what's the real problem here? Maybe the books are too heavy, or the straps are uncomfortable. Then, you'd Imagine different ideas, like a backpack with wheels or a super-light material that's strong enough to hold all your books. Next, you'd Plan by picking the best idea and sketching it out. Then comes the fun part—you get to Create it! Maybe you build a small model or try it out with different materials. Finally, you'd Improve it by testing it out, maybe asking your friends to try it and give you feedback.

But here's a secret: engineers don't always get it right the first time. In fact, they often have to try lots of different ideas before they find the best solution. And that's okay! Mistakes are just part of the process. Think about when you're building with blocks, and your tower keeps falling down. Do you give up? No way! You figure out what went wrong and try again. That's exactly how engineers think—they see every mistake as a chance to learn something new.

So, how do engineers solve problems? They break the problem down into smaller parts, think creatively, and aren't afraid to try, try again. The best part? You can use this way of thinking in your own life too. Whether it's figuring out a tricky math problem, coming up with a new game to play with friends, or finding a better way to organize your room, thinking like an engineer can help you solve any challenge that comes your way. So the next time you face a tough problem, don't get discouraged. Remember, you've got the skills of an engineer, and there's no problem you can't solve with a little creativity and determination!

Why Thinking Like an Engineer is Cool

Thinking like an engineer is all about seeing the world differently. You know how when you're playing with building blocks, you start to imagine what you can create—maybe a castle, a spaceship, or even a whole city? That's the first step to thinking like an engineer. It's about looking at something and thinking, "What can I make with this?" or "How can I make this better?" Engineers have this amazing ability to turn their ideas into real-life things that everyone can use.

Let's say you love animals and you notice that the birds in your neighborhood don't have enough places to nest. You could think like an engineer and come up with a plan to build birdhouses. First, you'd decide what materials to use, then you'd design the birdhouses to be just the right size. You might even add a special feature, like a little porch for the birds to perch on. Before you know it, you've created something that helps the birds and makes your neighborhood a better place. How cool is that?

Another reason thinking like an engineer is so awesome is because it makes you curious about how things work. Have you ever wondered how a roller coaster stays on the tracks or how your favorite video game was made? When you think like an engineer, you start to ask questions about everything around you. You want to know how things are built, why they work the way they do, and how they could be improved. This curiosity is like a superpower that helps you learn and discover new things every day.

And here's another secret: thinking like an engineer isn't just about building things. It's also about teamwork and sharing ideas. When engineers work on big projects, like designing a new car or building a bridge, they don't do it alone. They work with other engineers, each bringing their own ideas and skills to the table. It's like when you and your friends team up to build the ultimate fort—you share ideas, work together, and make something amazing. Thinking like an engineer teaches you how to work with others, listen to different ideas, and come up with the best solution together.

But perhaps the coolest thing about thinking like an engineer is that it helps you see challenges as opportunities. Instead of getting frustrated when something doesn't work, you start to see it as a puzzle to solve. Whether it's figuring out how to fix a broken toy, inventing a new game, or even finding a way to help others, thinking like an engineer gives you the confidence to tackle any challenge that comes your way. It's like having a superpower that helps you turn problems into possibilities.

So why is thinking like an engineer so cool? Because it lets you be creative, curious, and a great team player. It helps you solve problems, invent new things, and make the world a better place. And the best part? Anyone can think like an engineer—you don't need a special degree or a bunch of tools. All you need is your imagination, a love for learning, and a passion for creating.

Chapter 1

The Engineer's Toolbox

Welcome to Chapter 1, my friend! If you're going to think like an engineer, you're going to need some special tools. But don't worry—you won't need a big, heavy toolbox. Instead, the tools I'm talking about are ones you already have in your mind. Yep, your brain is the best toolbox you could ever ask for! In this chapter, we'll explore four of the most important tools engineers use every day: Curiosity, Creativity, Logic, and Teamwork. By the end of this chapter, you'll have some cool homework to try out and a step-by-step guide on how to get creative in your own life. Ready? Let's dive in!

Curiosity: Asking the Right Questions

Have you ever wondered why the sky is blue or how airplanes stay in the air? If so, congratulations! You've already started using the first tool in the engineer's toolbox: curiosity. Curiosity is all about asking questions and wanting to learn more about the world around you. It's like being a detective, but instead of solving crimes, you're solving the mysteries of how things work.

Engineers are naturally curious people. They look at everyday things and think, "How does that work?" or "Why was it made that way?" This curiosity drives them to explore, experiment, and discover new things. Imagine you're playing with a toy car. You might start to wonder, "What makes the wheels turn?" or "How does the car move faster when I push it?" That's your curiosity kicking in, and it's a powerful tool.

But curiosity doesn't stop at just asking questions. It's also about finding the answers. Let's say you're curious about how a flashlight works. You might take it apart (with permission, of course!) to see what's inside. Maybe you'll discover that there's a tiny bulb, some wires, and a battery. You might then wonder, "How do the wires and battery make the light turn on?" This kind of exploration is exactly what engineers do—they dig deeper to understand how things work.

Curiosity also helps you see problems as opportunities to learn something new. Imagine your bicycle has a flat tire. Instead of just getting frustrated, you could ask, "Why did the tire go flat?" and "How can I fix it?" Before you know it, you're learning how to repair a tire and maybe even thinking about how to design a tire that never goes flat.

Homework Time: For your curiosity homework, pick something around your house or school that you've always wondered about. It could be anything—a toaster, a lamp, or even a plant. Write down three questions you have about how it works or why it's made the way it is. Then, research or ask an adult to help you find the answers. If possible, try to take it apart and see for yourself how it works (but only if it's safe and you have permission!). Write down what you discover in a notebook. Congratulations, you've just taken your first step in thinking like an engineer!

Creativity: Finding New Ways to Solve Problems

Let's talk about creativity—the second tool in the engineer's toolbox. Have you ever built something out of

LEGO bricks that wasn't in the instruction manual? Maybe you created your own spaceship or designed a new kind of car. That's creativity in action! Creativity is all about thinking outside the box and finding new ways to solve problems.

Engineers use creativity every day to come up with innovative solutions to challenges. Let's say there's a town that needs a new bridge. The problem is, the river is really wide, and the weather is often stormy. A regular bridge might not be strong enough. So, an engineer would use their creativity to design a bridge that's not only strong but also able to withstand the stormy weather. Maybe they'd come up with a design that uses special materials or a shape that's never been tried before. That's the power of creativity!

Creativity is also about being open to new ideas and trying things that might seem a little unusual at first. Think about when you're playing a game and you come up with a new rule or a different way to play. That's your creativity shining through. Engineers do the same thing when they're faced with a problem that doesn't have an easy solution. They might look at the problem from a different angle, combine ideas from different fields, or even get inspiration from nature.

Here's a fun example: have you ever noticed how a gecko can climb walls and walk on ceilings? Engineers did! They studied the tiny hairs on a gecko's feet that let it stick to surfaces. Then, they used that idea to create a special kind of tape that sticks to anything, just like a gecko's feet. That's creativity at its best—finding inspiration in the world around you and using it to solve problems in new and exciting ways.

Homework Time: For your creativity homework, I want you to think about a problem you face every day. It could be something simple, like not having enough space in your backpack or a toy that keeps breaking. Now, imagine you're an engineer tasked with solving this problem. Draw or write down at least three different solutions, no matter how wild or unusual they might seem. Don't worry about whether they're realistic—just let your imagination run wild! Afterward, pick one idea and try to build a model or draw a detailed picture of it. Share your creation with a friend or family member and see what they think. Who knows? You might just come up with the next big invention!

Logic: Understanding How Things Work

Now, let's explore the third tool in our engineer's toolbox: logic. Logic is like the road map that helps you figure out how things work. It's all about understanding cause and effect—when you do one thing, what happens next?

Engineers use logic to solve problems in a step-by-step way, making sure that everything they design will work as planned.

Imagine you're playing a game of dominoes. You line up the pieces, and with one small push, they all fall down in a perfect line. That's logic in action. You know that if you push one domino, it will knock down the next one, and so on. Engineers think the same way when they're working on a project. They consider every step and make sure that each part of their design works with the others.

Logic is especially important when engineers are trying to figure out why something isn't working. Let's say you've built a robot, but it won't move. You'd use logic to troubleshoot the problem. First, you'd check if the batteries are charged. If they are, you might check the wires to see if they're connected properly. You'd keep testing and checking different parts until you figure out what's wrong. This logical thinking helps engineers solve problems efficiently and effectively.

But logic isn't just for fixing things— it's also for making things better. Let's say you love playing video games, but the game you're playing is too slow. Using logic, you might figure out that the problem is with the computer's memory. An engineer would use this knowledge to design a faster computer or to improve the game's software so it runs more smoothly. Logic helps you break down complex problems into smaller, manageable parts, making it easier to find a solution.

Homework Time

For your logic homework, I want you to create a simple Rube Goldberg machine. Now this is a machine that performs a very simple task in a complicated way, using a series of steps that trigger each other—like a domino effect. You could make a machine that pops a balloon, pours a glass of water, or turns off a light. Use objects you find around the house, like balls, ramps, and books. As you build, think about how each step will lead to the next. Write down the steps in order and test your machine. If something doesn't work, use logic to figure out why and make adjustments. Don't forget to have fun with it!

Working Together to Create Something Amazing

Last but not least, let's talk about teamwork—the fourth tool in our engineer's toolbox. Have you ever played a sport, worked on a school project with friends, or built a fort with your siblings? If so, you've already experienced the power of teamwork. Teamwork is all about working together with others to achieve a common goal.

Engineers rarely work alone. They collaborate with other engineers, scientists, designers, and even people from different fields. Each person brings their own skills and ideas to the table, making the project stronger and better. Imagine trying to build a giant sandcastle on the beach all by yourself. It would take forever! But if you work together with your friends, each person can take on a different task—one person gathers sand, another shapes the towers, and someone else decorates with shells. Before you know it, you've built an amazing sandcastle that's way better than anything you could have done alone.

Teamwork also means listening to others and being open to their ideas. Let's say you're designing a new playground with a group of friends. You might have a great idea for a slide, but your friend might have an even better idea for a climbing wall. By combining your ideas, you create a playground that everyone will love. Engineers do this all the time—they brainstorm together, share their thoughts, and build on each other's ideas.

But teamwork isn't just about sharing ideas—it's also about supporting each other. Sometimes, a project might hit a snag, and that's when teamwork really shines. If someone on the team is stuck, others can step in to help solve the problem. By working together, engineers can overcome challenges that might seem impossible if they were working alone.

Homework Time

For your teamwork homework, I want you to work with a friend or family member to create something together. It could be anything—a model of a building, a game, a piece of art, or even a simple invention. The important thing is that you both contribute ideas and work together to bring your creation to life.

Here's the step-by-step process for your teamwork project:

1. Brainstorm Together: Start by sitting down with your partner and coming up with a list of ideas. Don't be afraid to think big or outside the box!

Write down every idea, no matter how silly or wild it might seem.

2. Choose an Idea: Once you have a list, pick one idea that you both like the most. This should be something that excites both of you and that you're eager to work on.

3. Divide the Tasks: Now, figure out who will do what. Maybe one of you is great at drawing, while the other is better at building. Play to each other's strengths and divide the tasks in a way that feels fair and fun.

4. Work Together: As you start working, remember to communicate and help each other. If one of you gets stuck or isn't sure what to do

next, talk it through together. Teamwork is all about supporting each other and making sure everyone's ideas are heard.

5. Combine Your Work: Once you've both completed your tasks, bring everything together to see how it fits. Maybe you'll need to make some adjustments, and that's okay! Engineers often tweak their designs as they go along, especially when working in a team.

6. Test and Improve: If your project is something that can be tested (like a game or a model), try it out together. See how it works and if there's anything you can improve. Sometimes,

the best ideas come after you've seen your creation in action.

7. Celebrate Your Success: Finally, take a step back and admire what you've created together. Teamwork is all about combining your talents to create something amazing, and you've done just that!

After you've finished, share your project with others—show it to your family, friends, or even your teacher. Talk about how you worked together and what you learned from the experience. Teamwork isn't just a tool for engineers; it's something you'll use throughout your life, whether you're playing sports, doing school projects, or even just hanging out with friends.

And there you have it—four powerful tools that will help you think like an engineer: curiosity, creativity, logic, and teamwork. With these tools in your mental toolbox, you're ready to start solving problems, designing cool things, and making the world a better place. Remember, the more you use these tools, the better you'll get at thinking like an engineer.

So, what are you waiting for? Grab your curiosity, spark your creativity, apply your logic, and team up with others to start building something awesome today. The world is full of challenges just waiting for someone like you to solve them. And who knows? Maybe your next project will change the world!

Chapter 2

The Engineering Design Recipes

Now that you've learned about the tools engineers use, it's time to put those tools to work. In this chapter, we're going to dive into the Engineering Design Process, which is like a recipe that engineers follow to solve problems and create amazing things. And guess what? You can use this process to tackle your own challenges, whether you're at home, school, or anywhere else.

To make things fun and easy to understand, we're going to follow the story of a kid just like you—let's call him Alex—who uses the Engineering

Design Process to solve a problem he faces at school. Ready to get started? Let's go!

Step 1: Ask – Identifying the Problem

Every great invention starts with a problem. But before you can solve a problem, you need to figure out exactly what it is. This first step in the Engineering Design Process is called Ask.

Let's say Alex is having trouble with his backpack. Every day, he struggles to find his pencil case because it always ends up buried under all his books and lunchbox. The problem is clear: Alex needs a way to keep his pencil case easily accessible in his backpack.

In this step, you want to ask yourself some important questions:
- What exactly is the problem?
- Why is it a problem?
- Who does it affect?
- What do I want to achieve by solving it?

So, Alex asks himself, "Why is it hard to find my pencil case?" He realizes it's because everything in his backpack is jumbled together. He also asks, "What would make it easier to find?" Maybe a special pocket or a divider would help. By asking these questions, Alex identifies his problem:

How can I organize my backpack so I can find my pencil case quickly?

Homework Time: Think about a problem you face every day. Maybe it's something small, like not being able to keep your room tidy or having trouble remembering your homework assignments. Write down what the problem is, why it's a problem, and how you think it affects you or others. Once you've clearly identified the problem, you're ready to move on to the next step.

Step 2: Imagine – Brainstorming Solutions

Now that Alex knows what the problem is, it's time to let his imagination run wild. The second step in the Engineering Design Process is Imagine, where you brainstorm as many solutions as possible.

Alex sits down with a piece of paper and starts to imagine different ways to organize his backpack. He thinks about adding extra pockets, using a special case for his pencils, or even designing a backpack with see-through compartments so he can spot his pencil case right away.

The key here is not to limit yourself. There are no wrong answers during brainstorming! Alex knows that he might come up with some ideas that seem silly, but that's okay. Sometimes, the best solutions come from the craziest ideas.

Homework Time: Take the problem you identified in the first step and spend some time imagining different ways to solve it. Write down at least five different solutions, no matter how strange or unusual they might seem. Share your ideas with a friend or family member and ask them to add their own. The more ideas, the better!

Step 3: Plan – Making a Blueprint

After brainstorming, it's time to choose the best idea and make a plan. This is called Plan.

Alex decides that the best solution for his backpack problem is to create a simple divider that can separate his books from his pencil case. Now, he needs to figure out how to make it. He sketches a rough blueprint of what his divider will look like. He draws it with pockets for his pencils and a section for his books.

In this step, you want to think about what materials you'll need and the steps you'll take to create your solution. Alex lists the materials he'll need: some sturdy cardboard, scissors, tape, and maybe some fabric to make the divider look nice. He also writes down the steps:
1. Cut the cardboard to fit the backpack.

2. Attach the fabric to make pockets.
3. Insert the divider into the backpack.

Homework Time: Take one of the solutions you brainstormed and make a detailed plan. Draw a picture or diagram of your idea, and write down the materials you'll need and the steps you'll take to create it. If you're not sure how to start, ask an adult for help. Remember, a good plan is like a map that will guide you through the next steps!

Step 4: Create – Building Your Idea

Now comes the fun part—Create! This is where you get to build your solution.

Alex gathers all the materials he needs and starts to work. He carefully cuts the cardboard to the right size and tapes the fabric to create pockets. After a bit of cutting, folding, and taping, he has a divider that fits perfectly inside his backpack. It even has a special pocket just for his pencil case!

As you create your solution, remember to take your time and follow your plan. But don't worry if things don't go perfectly the first time—part of being an engineer is figuring out how to make things work as you go.

Homework Time: It's time to create your solution! Use the plan you made in the last step to build your idea. Take your time, and don't be afraid to ask for help if you need it. Once you're finished, test it out and see how it works. Does it solve the problem you identified? If not, don't worry—we're going to talk about testing and improving next!

Step 5: Test – Seeing if It Works

Once Alex finishes building his divider, it's time to put it to the test. This step is called Test.

The next morning, Alex loads up his backpack with all his school supplies. He's excited to see if the divider makes it easier to find his pencil case. As he packs, he realizes that his divider works great—it keeps everything organized, and he can grab his pencil case in seconds.

But testing doesn't always go smoothly. Sometimes, your solution might not work as planned, and that's okay! The important thing is to see what works and what doesn't. If something doesn't work, you can figure out why and think about how to fix it.

Homework Time: Test the solution you created. Use it in a real situation and see if it solves the problem. Pay close attention to what works well and what doesn't. Write down your observations in a notebook. Did your solution solve the problem? If not, what could you do to make it better?

Step 6: Improve – Making It Even Better

Even if Alex's divider worked well, there's always room for improvement. The final step in the Engineering Design Process is Improve.

After using the divider for a week, Alex notices that one of the pockets is a little too small for some of his pencils. He decides to make the pocket bigger and adds an extra section for his water bottle. By making these improvements, Alex's divider becomes even more useful.

Improving is all about taking what you've learned from testing and making your solution even better. Engineers are always looking for ways to improve their designs, even after they work. This step is what helps turn a good idea into a great one!

Chapter 3

Types of Engineers

Welcome to Chapter 3! Now that you've got the basics down about how engineers think and work, it's time to dive into the different types of engineers out there. You see, engineering isn't just one job—it's a whole world of exciting careers where people solve problems in their own unique ways. Let's explore some of these types of engineers together. You might just find something that sparks your interest!

Civil Engineers: Building Bridges and Roads

Have you ever crossed a bridge or traveled on a smooth highway and wondered who made it all possible? That's where civil engineers come in! Civil engineers are like the architects of our everyday world. They design, build, and maintain the structures that we use all the time, like bridges, roads, tunnels, and even the buildings we live and work in.

Imagine you're walking across a huge bridge that spans a wide river. Civil engineers had to figure out how to make that bridge strong enough to hold the weight of hundreds of cars, trucks, and people, all at the same time. They also had to make sure it wouldn't collapse in bad weather, like during a hurricane or earthquake. It's like putting together the world's most complicated puzzle, but the pieces are made of steel, concrete, and asphalt!

Let's take a look at an example. Imagine you and your friends are building a small bridge out of popsicle sticks for a school project. You need to think about how to make it strong, how to connect the sticks so they don't fall apart, and how to balance everything so your bridge doesn't tip over. That's a tiny version of what civil engineers do every day—but on a much bigger scale!

Homework Time: Try building your own mini bridge using things you have at home, like popsicle sticks, straws, or even LEGO bricks. Your challenge is to make a bridge that can hold the weight of a small toy car or a book. As you build, think like a civil engineer. What can you do to make your bridge stronger? After you're done, test it out and see how well it works!

Mechanical Engineers: Designing Machines and Gadgets

Next up, we have mechanical engineers. If you love taking things apart to see how they work, or if you've ever tried to build your own robot, then mechanical engineering might be right up your alley!

Mechanical engineers are the masterminds behind machines and gadgets. They design everything from the engines in cars to the gears in your bicycle, and even the robots used in factories. Their job is to figure out how to make things move, how to make them work efficiently, and how to make sure they don't break down.

Imagine you're trying to build a robot that can pick up toys from the floor. First, you'd need to think about what parts the robot needs, like arms to grab the toys and wheels to move around. Then, you'd need to figure out how to make those parts work together—how the gears should turn, how the motors should run, and how everything should fit inside the robot's body.

Mechanical engineers do this every day, but with much bigger projects. They might design a new type of car engine that uses less fuel, or a wind turbine that generates electricity from the wind. It's all about making things that move, work, and help us in our daily lives.

Homework Time: Try designing and building your own simple machine at home. It could be a small car made from cardboard and bottle caps, or a catapult that launches marshmallows. Think about how the different parts need to work together, and how you can make your machine move smoothly. Once you're done, test it out and see if it works as planned. If not, don't worry—just like a real engineer, you can always make improvements!

Electrical Engineers: Powering the World

Now, let's talk about the wizards of electricity—electrical engineers! These engineers are responsible for designing and creating the electrical systems that power our homes, gadgets, and even the lights on your street.

Have you ever wondered how flipping a switch can light up an entire room? That's thanks to electrical engineers. They design the circuits and systems that carry electricity from power plants all the way to your home. They also work on the tiny circuits inside your favorite devices, like video game consoles, smartphones, and computers.

Imagine you're trying to build a simple circuit to light up a small bulb. You'd need a power source (like a battery), some wires, and the bulb itself. You connect everything together, and when you complete the circuit, the bulb lights up! Electrical engineers do this on a much larger and more complex scale. They design everything from the power grids that supply electricity to entire cities to the tiny chips inside your electronics that make them run.

Electrical engineering is all about understanding how electricity works and how to control it. It's like learning a new language—the language of circuits and currents!

Homework Time: Try building a simple circuit at home. You can use a battery, some wires, and a small light bulb or an LED. If you don't have these materials, you can also find online circuit simulation tools to play around with. See if you can make the light bulb turn on and off by creating a switch in your circuit. As you work, think about how electrical engineers design circuits to power everything from light bulbs to computers.

Environmental Engineers: Protecting Our Planet

As we become more aware of the impact humans have on the environment, the role of environmental engineers is more important than ever. These engineers work to protect our planet by finding ways to reduce pollution, manage waste, and conserve natural resources like water and energy.

Imagine you're at a park, and you see trash scattered everywhere. An environmental engineer's job might be to come up with a better way to manage waste so that trash doesn't end up in our parks, rivers, or oceans. They might design systems to clean polluted water, create ways to recycle materials more efficiently, or develop new forms of renewable energy like solar or wind power.

Environmental engineers also think about the future. They work on projects that help communities prepare for challenges like climate change, ensuring that we have clean air, water, and energy for generations to come.

For example, let's say you want to help clean up your local park. You could start by organizing a group to pick up trash, but an environmental engineer might go a step further. They'd think about how to prevent trash from being left there in the first place, maybe by designing better trash bins or creating educational programs to teach people about the importance of keeping the environment clean.

Homework Time: Take a walk around your neighborhood or a nearby park and look for ways you could help protect the environment. Maybe you'll notice areas that need more recycling bins or places where water is being wasted. Write down your observations and come up with a plan to address one of these issues. You could even try to implement your plan with the help of friends, family, or your school!

Aerospace Engineers: Reaching for the Stars

Last but not least, let's talk about aerospace engineers—the people who design everything that flies, from airplanes to rockets to spacecraft! If you've ever dreamed of traveling to space or flying a plane, aerospace engineering might be the field for you.

Aerospace engineers are responsible for designing and testing all kinds of flying machines. They work on airplanes that carry passengers across the world, rockets that send satellites into orbit, and spacecraft that explore distant planets. Their job is to make sure these machines are safe, efficient, and capable of amazing feats like traveling at super-fast speeds or surviving the harsh conditions of space.

Imagine you're building a model rocket. You'd need to think about how to make it aerodynamic (so it can fly through the air smoothly), how to balance the weight so it doesn't tip over, and how to power it so it can launch into the sky. Aerospace engineers do this on a much larger scale, designing rockets that can travel to the moon or even Mars!

Aerospace engineering is all about pushing the boundaries of what's possible. It's a field where you can literally reach for the stars!

Homework Time: Try building your own paper airplane or model rocket. Experiment with different designs to see which one flies the farthest or the fastest. Think about how you can make your design more aerodynamic by adjusting the shape or adding wings. Once you're happy with your design, test it out and see how well it flies. If it doesn't work as well as you'd like, don't worry—just like aerospace engineers, you can keep refining your design until it's ready for takeoff!

Chapter 4

Real-Life Engineering Challenges

Hey there, future engineers! Ready to get hands-on and tackle some real-life engineering challenges? In this chapter, we're going to dive into some fun and creative projects that will help you think like an engineer. Each challenge will give you a taste of what it's like to solve problems, design solutions, and test your ideas—just like a real engineer would. So, roll up your sleeves and let's get started!

Building a Strong Bridge with Spaghetti and Marshmallows

First up, let's try building a bridge—but not just any bridge. We're going to build one using spaghetti and marshmallows! Sounds a little silly, right? But this challenge is a great way to learn about structures, balance, and strength.

Step 1: The Challenge
Imagine you need to build a bridge that can hold up a small object, like a toy car or a book. Your materials are uncooked spaghetti (for the beams) and marshmallows (for the joints). How do you think you can make your bridge strong enough to hold the weight without collapsing?

Step 2: The Design

Start by thinking about how real bridges are built. They often have triangles in their design because triangles are super strong shapes. Triangles help distribute weight evenly, which keeps the bridge from bending or breaking. So, as you start building your spaghetti bridge, try using triangles in your design.

Step 3: The Build
Now it's time to start building! Take your spaghetti and marshmallows and begin constructing your bridge. Connect the spaghetti sticks with marshmallows to form triangles. Remember to be gentle—spaghetti can be pretty fragile, and you don't want it to break before your bridge is even done!

Step 4: The Test

Once your bridge is built, it's time to see how strong it is. Place your small object in the middle of the bridge and see if it holds up. If it does, awesome job! If it doesn't, that's okay too—just like a real engineer, you can go back and make improvements.

Creating a Simple Circuit with Batteries and Bulbs

Next, let's play around with electricity! Don't worry—it's not as scary as it sounds. We're going to build a simple circuit to light up a small bulb using just a few basic materials.

Step 1: The Challenge

Your goal is to create a circuit that can turn on a light bulb using a battery, some wires, and the bulb. But first, do you know what a circuit is? A circuit is like a path that electricity travels along. For the bulb to light up, the circuit needs to be complete, meaning there can't be any gaps in the path.

Step 2: The Design

Think about how you're going to connect everything. The battery has two ends—a positive (+) and a negative (-). The electricity needs to flow from the positive end of the battery, through the wire, into the bulb, and back to the negative end of the battery. If there's a break in this path, the electricity won't flow, and the bulb won't light up.

Step 3: The Build
Start by connecting one wire to the positive end of the battery and the other end of the wire to the bottom of the bulb. Then, connect another wire to the side of the bulb and attach the other end of this wire to the negative end of the battery. Once everything is connected, the bulb should light up!

Step 4: The Test
Now, test your circuit by seeing if the bulb lights up. If it does, congratulations—you've just created your first circuit! If it doesn't work,

double-check your connections and make sure the wires are securely attached to the battery and the bulb. Sometimes, even a tiny gap can stop the electricity from flowing.

Designing a Paper Airplane that Flies the Farthest

Who doesn't love making paper airplanes? But this time, we're going to think like aerospace engineers and design a plane that can fly as far as possible.

Step 1: The Challenge
Your goal is to design a paper airplane that can fly farther than anyone else's. To do this, you'll need to think about the shape and weight of your

airplane. What kind of design do you think will help it fly farther?

Step 2: The Design
Start by thinking about the wings. Bigger wings might help your plane glide farther, but they also add weight, which could slow it down. You'll also want to think about the body of the plane. Should it be long and thin, or short and wide? These are all things you can experiment with.

Step 3: The Build
Fold your paper into the design you've chosen. Try to make your folds as sharp and straight as possible—this will help your airplane fly smoothly. You can also experiment with adding a little weight to the front of the plane

by folding the tip over or by adding a paper clip. This can sometimes help the plane fly farther.

Step 4: The Test
Now it's time to fly your airplane! Find an open space, like a backyard or a park, and see how far your plane goes. If it doesn't fly as far as you'd like, go back and make some changes. Maybe try a different wing design or adjust the weight. Keep testing and improving until you get the best result.

Inventing a Device to Clean Up a Messy Room

Now for a fun challenge—let's invent something that can help clean up a messy room. Wouldn't it be awesome to have a device that could put away toys, books, and clothes all by itself?

Step 1: The Challenge

Your mission is to come up with an idea for a device that can help clean up your room. It could be something as simple as a grabber arm that picks up toys or as complex as a robot that sorts everything into its place. Use your imagination—there are no limits here!

Step 2: The Design

Start by thinking about what your device needs to do. Does it need to move around the room? How will it pick things up? Will it need to sort items into different places? Once you have a clear idea, draw a picture of your device and label the different parts.

Step 3: The Build
If you have materials at home, like cardboard, string, or old toys, try building a simple model of your device. It doesn't have to work perfectly—this is just to help you think through your idea. You could even use LEGO bricks to create a basic version of your invention.

Step 4: The Test
Once you've built your model, test it out! See if it can help with any small tasks, like picking up a toy or moving an object from one place to another. If it works, great! If not, think about

what you could change to make it better.

Homework: Engineering Challenge Recap

For this chapter's homework, I want you to pick one of the challenges above and dive even deeper. Here's what you should do:

1. Choose Your Challenge: Pick the challenge that excites you the most—whether it's building a bridge, creating a circuit, designing a paper airplane, or inventing a cleaning device.

2. Make a Plan: Write down or draw your plan for how you're going to

tackle the challenge. Think about what materials you'll need, how you'll put everything together, and what you hope the end result will be.

3. Build and Test: Start building your project! Once it's built, test it out and see how well it works. Remember, it's okay if it doesn't work perfectly the first time. Engineers often have to test and improve their designs multiple times.

4. Reflect and Improve: After testing, think about what went well and what didn't. Write down any improvements you'd make if you were to try the challenge again.

5. Share Your Work: If you can, share your project with friends, family, or even your teacher. Explain how you thought like an engineer to solve the problem, and what you learned from the experience.

By the end of this homework, you'll have a deeper understanding of what it's like to think like an engineer—and you might even have some cool new creations to show off! Keep practicing, keep experimenting, and remember that every mistake is just another step towards success. Happy engineering!

Chapter 5

Becoming an Engineer

Hey, future engineers! Now that you've learned a lot about what engineers do, how they solve problems, and why thinking like an engineer is super cool, you might be wondering: How can I become an engineer too? Well, you're in luck! In this chapter, we're going to talk about the steps you can take right now to start your engineering journey. It's never too early to start learning, practicing, and having fun with engineering!

What You Can Do Now

Becoming an engineer doesn't happen overnight—it takes practice, learning, and a lot of curiosity. But guess what? You can start right now, even while you're still in school! There are so many things you can do to build your skills and knowledge, and the best part is, most of them are really fun.

1. Get Hands-On with Building Projects

One of the best ways to start thinking like an engineer is to get hands-on with building projects. Remember those challenges we did in the last chapter? Keep doing things like that! You can build models, design machines with LEGO, or even create your own inventions out of cardboard and tape. The more you build, the more you'll learn about how things work—and the more confident you'll become in your abilities.

2. Explore Science and Math

Engineers use a lot of science and math in their work, so it's important to start getting comfortable with these subjects now. But don't worry—science and math can be really fun! Try conducting simple experiments at home, like mixing baking soda and vinegar to see what happens, or measuring how far your paper airplane can fly. You can also play math games online or try solving puzzles that challenge your brain. The more you explore, the more you'll discover how cool these subjects really are.

3. Read Books and Watch Videos about Engineering

There are tons of books, videos, and even TV shows that are all about engineering. Check out books from your library that talk about famous engineers or cool inventions. Watch videos on YouTube that show how things are made, like airplanes, robots, or even toys. The more you learn about what engineers do, the more inspired you'll be to create your own projects.

4. Practice Problem-Solving

Without overstating it, Engineers are amazing problem-solvers, so start practicing now! You can do this by playing games that challenge your brain, like puzzles or strategy games. You can also look around your home or school for problems that need solving. Maybe you can design a new way to organize your books or figure out how to make your morning routine faster. Every time you solve a problem, you're thinking like an engineer!

5. Keep a Journal of Your Ideas

Engineers are always coming up with new ideas, so start keeping a journal where you can write down your own.

Whenever you think of a cool invention or a better way to do something, jot it down. You can also draw pictures of your ideas to help you remember them. Over time, you'll have a whole book full of your amazing engineering ideas!

How to Stay Curious and Keep Learning

Curiosity is one of the most important qualities an engineer can have. It's what drives engineers to ask questions, explore new ideas, and never stop learning. So, how can you stay curious and keep that spark of curiosity burning bright?

1. Ask Questions About Everything

The world is full of fascinating things, and the best way to learn about them is by asking questions. How do airplanes fly? Why does the sun rise and set? How does a computer know what to do when you press a button? Every time you ask a question, you're opening the door to learning something new. Don't be afraid to ask lots of questions—you might be surprised by what you discover!

2. Try New Things

One of the best ways to keep learning is by trying new things. Maybe you've never built a model car before, or you've never experimented with coding. Give it a try! Even if it's something you've never done before, it's a great opportunity to learn something new. Plus, trying new things can be a lot of fun, and you might discover a new hobby you really enjoy.

3. Visit Museums and Science Centers

If you ever have the chance to visit a museum or science center, take it! These places are full of hands-on exhibits and interactive displays that let you explore science, technology, and engineering in a fun and engaging way. You might get to build a bridge, explore the stars in a planetarium, or even see robots in action. It's a great way to spark your curiosity and learn more about the world around you.

4. Stay Up-to-Date with New Technology

Technology is always changing and evolving, and it's important to stay up-to-date with the latest trends. Read articles, watch videos, and follow the news to learn about new inventions, discoveries, and advancements in engineering. The more you know about what's happening in the world of technology, the more prepared you'll be to tackle new challenges and come up with innovative ideas.

5. Never Stop Wondering

Finally, the most important thing you can do to stay curious is to never stop wondering. Always keep that sense of wonder and amazement about the world. Whether you're looking up at the stars, watching a bird build a nest, or trying to figure out how your favorite toy works, keep asking questions and keep exploring. The more you wonder, the more you'll learn—and the more you'll grow as an engineer.

Cool Engineering Clubs and Competitions

Did you know that there are clubs and competitions just for kids who love engineering? Joining a club or entering a competition is a great way to meet other kids who share your interests, learn new skills, and have a ton of fun!

1. Join a Robotics Club

Robotics clubs are super popular, and for good reason—they're awesome! In a robotics club, you'll get to build and program robots that can do all sorts of cool things, like navigating mazes, picking up objects, or even playing games. You'll also get to work with a team, which is a great way to practice your teamwork skills. Plus, many robotics clubs enter competitions where you can show off your robot's skills and compete against other teams.

2. Participate in Science Fairs

Science fairs are a classic way to show off your engineering skills. You can enter a project that demonstrates a scientific concept, solves a problem, or showcases a cool invention. Science fairs are a great way to challenge yourself, learn more about a topic you're interested in, and maybe even win a prize!

3. Enter Engineering Challenges

There are all sorts of engineering challenges and competitions out there, from bridge-building contests to coding challenges. These competitions are designed to test your problem-solving skills and creativity, and they're a lot of fun. You can enter as an individual or as part of a team, and you'll get to see how your ideas stack up against others.

4. Join a Makerspace

Makerspaces are places where people can come together to create, invent, and learn. They're full of tools and materials for building all sorts of things, from 3D-printed models to electronic gadgets. Many makerspaces offer classes and workshops where you can learn new skills and work on projects with other kids. It's a great way to get hands-on experience with engineering and technology.

5. Start Your Own Club or Group
If there isn't an engineering club or competition in your area, why not start your own? You can gather a group of friends who are interested in engineering and start meeting regularly to work on projects, share ideas, and learn from each other. You

could even organize your own mini-competitions or challenges. Starting your own club is a great way to take charge of your learning and have fun with your friends at the same time.

Homework

Your Engineering Action Plan

For this chapter's homework, I want you to create your own engineering action plan. Here's what you need to do:

1. Set a Goal: Think about what you want to achieve as an engineer. Do you want to learn more about robotics? Build your own inventions? Enter a competition? Write down your goal so you can keep it in mind as you work on your action plan.

2. Make a List of Activities: List out the activities you can do to reach your goal. For example, if you want to learn

more about robotics, you might join a robotics club, watch online tutorials, or build your own robot at home. Be specific and include as many activities as you can think of.

3. Create a Timeline: Set a timeline for when you want to complete each activity. It could be a week, a month, or even longer—whatever works best for you. Make sure your timeline is realistic and gives you enough time to complete each activity without feeling rushed.

4. Find Resources: Think about what resources you'll need to complete your activities. Do you need to borrow books from the library, buy materials, or sign up for a class? Make a list of

the resources you'll need and start gathering them.

5. Share Your Plan: Finally, share your action plan with someone who can help you stay on track, like a parent, teacher, or friend. Explain what you want to achieve and how you plan to do it. They might even have some helpful suggestions or ideas to add to your plan.

By the end of this homework, you'll have a clear plan for how you can start your journey to becoming an engineer. Remember, the most important thing is to keep learning, stay curious, and have fun! The world needs creative thinkers like you to solve problems, invent new things, and make the world a better place. Happy engineering!

Chapter 6

Wrapping It Up: You're an Engineer

Wow, can you believe how far you've come? We've explored so much together in this book, and now, it's time to wrap it all up. But before we say goodbye, let's take a moment to reflect on everything you've learned, talk about how to keep thinking like an engineer, and explore where you can go from here. You're not just a reader anymore—you're an engineer in the making!

Reflecting on What You've Learned

Let's take a quick trip down memory lane. Do you remember when we started this journey? You were curious about what engineers do, and now, look at you—you've learned about all kinds of engineers, from those who build bridges and roads to those who send rockets into space. You've even tackled some real-life engineering challenges and started thinking about how to solve problems like a pro.

Let's do a quick recap:

- You learned that engineers are problem-solvers. They use creativity, logic, and teamwork to come up with solutions that make our lives better. Whether it's designing a safer car, building a taller skyscraper, or inventing a new gadget, engineers are always thinking about how to improve the world.

- You explored different types of engineers. Civil engineers, mechanical engineers, electrical engineers, environmental engineers, aerospace engineers—the list goes on and on! Each type of engineer has a special role, but they all have one thing in common: they love solving problems and making things work.

- You got hands-on with engineering challenges. Remember building a strong bridge out of spaghetti and marshmallows? Or designing a paper airplane that flies the farthest? These challenges weren't just fun—they also taught you important skills like planning, testing, and improving your designs.

- You discovered the engineering design process. From identifying the problem to brainstorming solutions, making a plan, building your idea, testing it, and then making it even better—you've learned the steps that engineers use every day to bring their ideas to life.

- You started thinking about your future as an engineer. Whether you're interested in joining a robotics club, entering a science fair, or just tinkering with your own inventions at home, you're already on your way to becoming an engineer. And the best part? You can keep learning and practicing, no matter where you are.

Now that we've reflected on all you've learned, let's talk about how you can keep thinking like an engineer, even after you close this book.

How to Keep Thinking Like an Engineer

Just because you've finished this book doesn't mean your engineering journey is over—in fact, it's only just beginning! Here are some tips to help you keep thinking like an engineer in your everyday life:

1. Keep Asking Questions

Remember when we talked about the power of curiosity? Engineers are always asking questions: How does this work? Why does that happen? What if we tried this instead? Keep asking questions about the world around you, and don't be afraid to explore the answers. The more you ask, the more you'll learn.

2. Practice Problem-Solving

Problems are everywhere, and that's a good thing for an engineer like you! The next time you face a challenge, whether it's fixing a broken toy or figuring out how to organize your room, approach it like an engineer. Identify the problem, brainstorm solutions, make a plan, and test your ideas. Every problem you solve helps you sharpen your engineering skills.

3. Get Creative

Engineers are creative thinkers—they're always coming up with new ideas and finding innovative ways to solve problems. So, keep flexing those creative muscles! Try building something new out of materials you have at home, or design a solution to a problem you see in your community. The more you practice creativity, the better you'll get at thinking outside the box.

4. Work with Others

Teamwork is a big part of engineering. Whether you're working on a school project, participating in a club, or just playing with friends, practice working together to solve problems. Listen to others' ideas, share your own, and find ways to combine your strengths. Remember, great things happen when we work together!

5. Keep Learning

The world of engineering is always changing, with new technologies and ideas popping up all the time. Keep learning by reading books, watching videos, and exploring new topics. If something sparks your interest—like robotics, coding, or renewable energy—dive in and learn as much as you can. The more you know, the more powerful your engineering brain will become!

6. Don't Be Afraid to Fail

One of the most important lessons in engineering is that failure is okay. In fact, it's a key part of the learning process. If something doesn't work the way you planned, don't get discouraged. Instead, think about what went wrong, make some changes, and try again. Every failure brings you one step closer to success.

Where to Go from Here

So, what's next for you, future engineer? The possibilities are endless! Here are a few ideas for where you can go from here:

1. Start a New Project

Now that you've learned so much about engineering, why not start a new project? It could be something simple, like building a model of a bridge, or something more complex, like designing your own invention. Whatever you choose, make sure it's something you're excited about. The more passionate you are, the more fun you'll have—and the more you'll learn.

2. Join a Club or Competition

If you haven't already, consider joining a club or entering a competition. There are so many options out there, from robotics clubs to science fairs to engineering challenges. Not only will you get to work on cool projects, but you'll also meet other kids who share your interests. Plus, competitions are a great way to test your skills and see how your ideas stack up against others.

3. Explore Engineering Careers

If you're curious about what it's like to be an engineer, start exploring different engineering careers. You can read about what different types of engineers do, watch videos of engineers in action, or even reach out to an engineer in your community to ask them about their job. The more you learn about engineering careers, the better you'll understand what kind of engineer you might want to be.

4. Keep a Journal of Your Ideas

As you continue on your engineering journey, keep a journal where you can jot down your ideas, sketches, and thoughts. This journal will be like a treasure chest of your creativity, full of ideas that you can come back to and develop over time. Plus, it's a great way to see how much you've grown as an engineer.

5. Stay Curious and Keep Exploring

Finally, remember to stay curious and keep exploring the world around you. Whether you're learning about new technologies, experimenting with new materials, or just asking questions about how things work, keep that sense of wonder alive. The world is full of amazing things waiting to be discovered—and with your engineering mindset, you're ready to explore them all!

Homework: Your Engineering Reflection

For your final homework, I want you to take some time to reflect on everything you've learned and think about what you want to do next. Here's what you need to do:

1. Write About Your Favorite Part of the Book: Think back on everything we've covered in this book. What was your favorite part? Did you enjoy learning about different types of engineers, tackling the challenges, or something else? Write a paragraph or two about what you enjoyed the most and why.

2. Set a New Engineering Goal: Now that you've learned so much, it's time to set a new goal. What do you want to achieve next? It could be anything

from building a new project to learning more about a specific type of engineering. Write down your goal and think about what steps you can take to reach it.

3. Share Your Ideas with Someone: Talk to a parent, teacher, or friend about what you've learned and what you want to do next. Share your favorite parts of the book, your new goal, and any ideas you have for future projects. Getting feedback and support from others is a great way to stay motivated and inspired.

4. Start Your Next Project: Finally, take the first step toward your new goal by starting a new project. It doesn't have to be big—just something that gets you thinking like an

engineer. Remember, every project you work on helps you build your skills and knowledge.

By completing this homework, you'll be well on your way to becoming an even more amazing engineer. Remember, the journey doesn't end here—there's a whole world of engineering waiting for you to explore. So keep building, keep learning, and keep dreaming big. The world needs engineers like you to solve problems, create new technologies, and make the world a better place. You've got this, future engineer!

www.ingramcontent.com/pod-product-compliance
Lightning Source LLC
Chambersburg PA
CBHW071045240526
45471CB00014B/581